To my daughters
Mackenzie and Sydney,
whose interest in jobs
inspired this book.

- Doug Kerwin

Copyright © 2021 by Doug Kerwin

All rights reserved. This book or any portion thereof may not be reproduced or used in any manner whatsoever without the express written permission of the publisher except for the use of brief quotations in a book review.

Author:	Illustrator:	Editor:
Doug Kerwin	Irfan Assaat	Crystal Watanabe
www.dougkerwin.com	www.picudesign.com	www.pikkoshouse.com
dwkerwin@gmail.com	irfan.assaat@gmail.com	crystal@pikkoshouse.com

ISBN 979-8-9854930-0-9

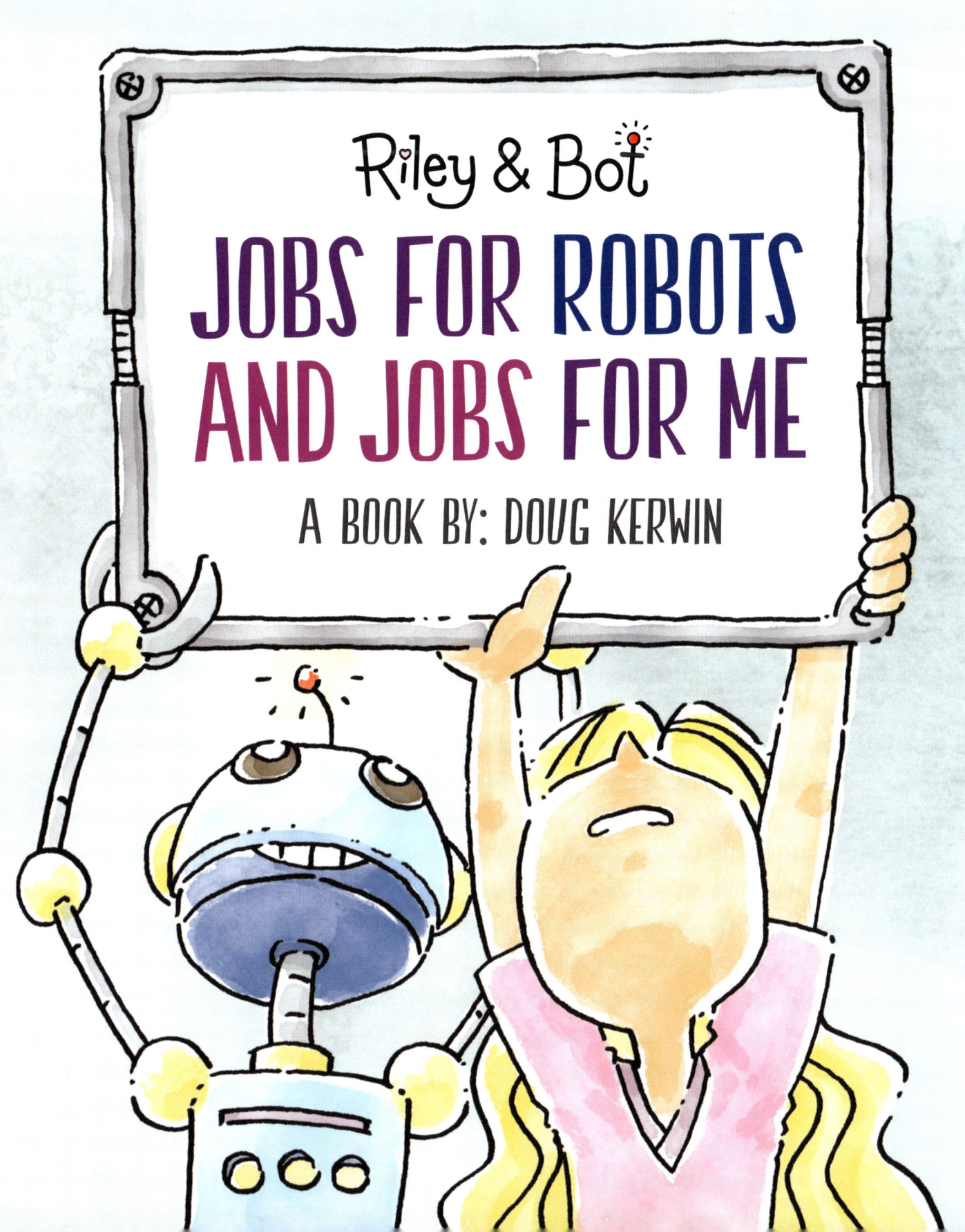

Just a few years from now,
in a house not far away,
was a girl on her couch,
just enjoying her day.

Next to her was her best friend, Bot.
She read her homework to Bot out loud.
"'What do you want to be when you grow up?'
How about something to make my parents proud?"

"My name is Riley, and I'm seven years old…
But I don't know what I want to be,
truth be told."

"I might want to be a fireman or a stylist who cuts hair.

Or a chef, a doctor, a driver, or a designer with flair!"

Bot, I need some help, because I really can't decide...
"BEEP! Of course. *BEEP!"* the robot replied.

"You know, Riley, not all of those jobs you named need to be done by people," the robot explained.

"Oh, Bot, surely you don't think that's true!

Remember my mom's friend, the photographer named Sue?

And don't forget about our neighbor, Fireman Lou."

"Oh, yes," said Bot, "robots and computers can do some of that work."
"No way!" Riley replied with a smirk.

"BEEP! Yes, that is certainly true!
Why don't you ask me about some jobs?
Maybe we can find one that's right for you."

"What about the job that my uncle Pete has? He drives a big truck all day and listens to jazz."

"Well," said Bot, "robots can do that job too. They don't need to stop like regular people do.

"They can drive all day and all night and never get tired. Robots don't need sleep, and that's why they get hired."

"Not just trucks, but taxis too.

Robots don't always need a body with arms and legs to do what you do.

They don't even need a seat, so there's more room for you!"

"What about an accountant like my aunt Stace?
Surely that's a job that a robot can't replace."

"Don't you know?" said Bot.
"Math is a robot's middle name!
They can crunch numbers so fast,
it's their claim to fame."

"And robots can do factory work all night and all day. Robots are stronger, work longer, and never delay.

They don't get sick, and they work hard come what may."

"A doctor!
They have to be skilled and super smart too.
I bet that's a really hard job
that even robots can't do."

"Actually, robots help doctors in hundreds of ways.
They can see what's making you sick
with almost no delays.

There are robots that can operate on your arm.
A robot's steady hand keeps you safe from harm."

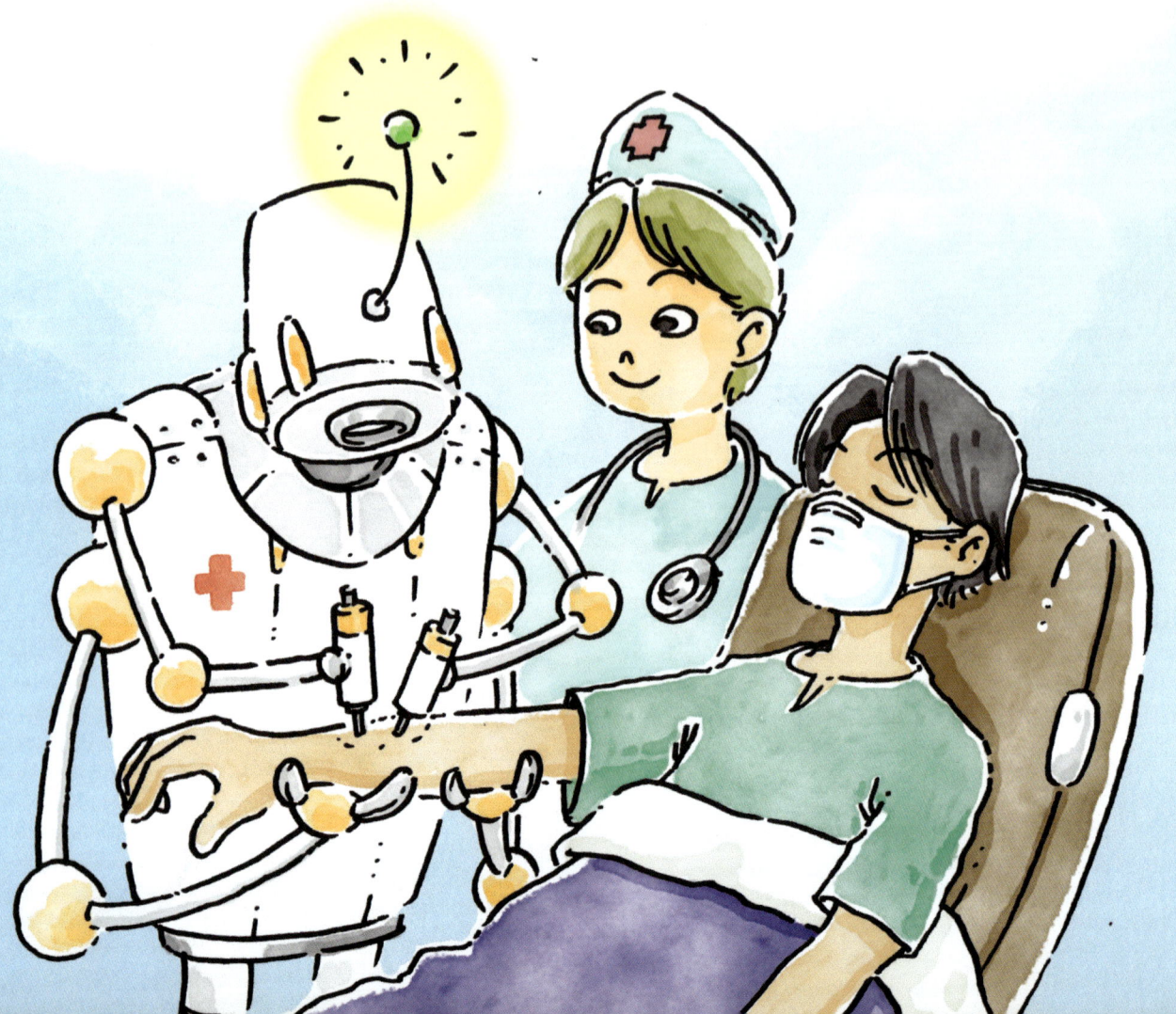

"Well, Bot, here's another
question that I must ask of you too.
Do the robots ever get paid for
all of those jobs that they do?"

"No, I'm afraid, when robots work,
they don't get paid.

People who own the robots get the money instead,
They'll make a lot of money so they'll really be ahead."

"Riley, I'd suggest", the robot said,
"that you make your own robots,
now that's using your head."

"*BEEP!*" said the robot.
"That's simply not true.
There are a lot of things that we robots cannot do.

There are many skills that we robots lack.
Here are a few careers that
robots cannot hack."

Let's start with complicated jobs.
Like people who run businesses
while the robots push buttons and turn knobs.

They have the gift of grand visions
Helping guide the company
and making the big decisions.

Or engineers who write computer programs in tech. They shape our digital future, keeping robots in check.

And there are jobs requiring
compassion and trust.
That's certainly a field that
leaves us robots in the dust.

A nurse, a nanny, or a therapist.
All of those jobs should be on your list.

Robots are great with numbers
and labor that's hard on your back,
But a true emotional connection
is something we lack.

And now that I've got your attention,
The future of jobs lies with invention.
There will be jobs that
we don't even know of yet.
But they will be fun and innovative,
on that you can bet!

Jobs like a Yambit Optimizer
or a Jimbit Joiner.
Or a Transit Tokenizer
or a Pricket Pointer.

You could be a
Jeffron Steffron
Babillion Mounter.
Or perhaps a Hebron Chevron
Civilian Counter.

Imagine how much fun
those jobs will be!
Allowing you to maximize
your creativity.

"You've given me so much to think about," Riley said, as she finished her homework and headed off to bed, with a mixture of worries and excitement about what lies ahead.

For Parents

People need a job, a purpose. Work is a fundamental part of being human, and even young children delight in considering what kind of work they might want to do in their life. The future of work is on course to undergo change unlike the world has ever seen before. As a technologist and an entrepreneur, the future looks very exciting to me—it's a world I want to live in. At the same time, it's scary, and I worry about my kids and the opportunities they might have.

I used the word "robots" throughout this book to make it more relatable to children, but really much of this is AI (artificial intelligence). Many people seem to underestimate the impact and timeline that AI will have on jobs and our economy. It's hard enough to believe that autonomous vehicles aren't far away from displacing millions of transportation workers. It's even more difficult to think that you might go through years of graduate school only to find that your area of discipline is performed better by AI. It's a tough message. There is so much change coming so fast that it is hard for anyone to believe.

While there will be new jobs created that don't exist yet, AI will be better than us at many tasks, and the owners of the AI are set to receive huge rewards while the rest of us may struggle to shift into work still better performed by people. We must reinvent ourselves as the landscape of work is reinvented before our eyes.

I wanted to encourage children and parents to start thinking and talking about this change, and to not be left behind. They say to keep opinions out of children's books. There are some loaded concepts here, and a lot of uncertainty as to how things will play out, however, I think children deserve a glimpse of a possible world they may find themselves in when they are old enough to get a job.

It's harder to envision the jobs that are created than it is to envision the jobs that are eliminated. The writing is on the wall for many industries that are ripe to become automated by AI, so it is easy enough to guess the jobs that will be lost to that automation. But there will be new jobs created as well. I don't pretend to know what those are, but I have faith that we will find new areas of work which we can't fathom today, and that this work will likely be done side by side with robots and AI.

HOMEWORK ASSIGNMENT

NAME:

DATE:

What do you want to be when you grow up?

List 1 or 2 jobs and describe what you like about the job and why you think it's a good choice. Use complete sentences and correct punctuation.

Printed in Great Britain
by Amazon